BEYOND CUT

REAL STORIES OF
REAL FREEDOM
WITH BONUS STUDY GUIDE

NANCY ALCORN

WinePress Publishing (PO Box 428, Enumclaw, WA 98022) functions only as book publisher. As such, the ultimate design, content, editorial accuracy, and views expressed or implied in this work are those of the author.

ISBN 13: 978-1-57921-935-2
ISBN 10: 1-57921-935-7
Library of Congress Catalog Card Number: 2007937888

To those who are ***desperate*** *for help*
But feel there is no hope,
This book has been placed in your ***hands*** *for a reason—*
It is no accident that you are reading this even now.
My ***prayer*** *is that you will read on,*
Because this book was written for you.
If you receive this message,
You will ***never*** *be the same!*

—Nancy Alcorn

CONTENTS

Introduction 1

Chapter One: Alyssa's Story—My Secret 3

Chapter Two: Bethany's Story—A Perfect Life 11

Chapter Three: Heather's Story—The Monster in the Mirror 19

Chapter Four: Jessica's Story—Consumed with Lies 27

Chapter Five: Katie's Story—A Silent Statistic 35

Chapter Six: Natalie's Story—I Made a Choice 43

Chapter Seven: Tiffany's Story—A Way of Escape 51

Chapter Eight: Valorie's Story—A Vicious Cycle 59

About Mercy Ministries 67

About the Author 69

INTRODUCTION

After reading Cut: Mercy for Self-Harm, you should have a better understanding of what self-harm is, and know how you can break free and stay free from this life-controlling issue.

Self-harm is often an expression of deep pain that is communicated through self-destructive behaviors. It may be a secret that you have never shared with anyone, but you are not alone in your struggle. Each chapter in this book includes a story of a young woman who fell into the cycle of self-harm, but found the way out and is now living a life of freedom. You, too, can experience freedom as you apply the same principles to your own life.

At the end of each story, you will find practical examples of how to apply these truths to your life. You will also learn how to communicate and express your emotions in a healthy way. Working through the questions at the end of each chapter will help you get to the root cause of the issues behind your destructive behaviors. I encourage you to go over your answers to each question with a trusted friend or mentor. Allow someone you trust to guide you on the path of freedom and answer any questions that you might have along the way.

As you take the next step to break free from the bondage of self-harm, pray that God will open your heart so you can identify the root issues at hand. Let God restore hope to your life as you embrace true and lasting freedom from self-harm.

Chapter One

ALYSSA'S STORY: MY SECRET

I *had* to be punished for any wrong I did. This was my secret.

I believed I was worthless and deserved to be punished, so I harmed myself. Self-injury was my way to escape the painful, scary, and shameful emotions that were stirred up by the sexual abuse I had experienced beginning when I was five years old. My abusers would hurt me while they were sexually abusing me, so I began associating the emotions I felt while I was being abused with physical pain. Just like taking a drink when you are thirsty, it was ingrained in my mind to hurt myself whenever I felt guilty or ashamed. I found relief and release for a fleeting moment when I cut myself.

Since no one knew my secret, it didn't seem to affect the few relationships I had. But in order to protect my secret, I felt the need to remain closed, keeping people at a distance. I didn't trust anyone. Therefore, I avoided relationships at all costs. I experienced an endless cycle of shame that fed the lies: "I'm worthless. Others hate me. I can't be forgiven." As the lies continued to swarm in my head, the urge to cut myself became overwhelming.

I never imagined anyone could forgive me, especially God. Even though the Bible says that anyone who believes in Jesus and asks for forgiveness can have a relationship with Him, I was convinced I was an exception. I hated the person I had become and despised the life in which I was trapped. I thought I was doing the right thing by hurting myself.

While growing up, I went to a different church every week with my parents. One week it would be a church where I was taught to worship God, and the next week it would be a completely opposite kind of gathering that encouraged me to worship Satan. I was confused about God and wondered what He was really like. I became very frustrated trying to understand God. As I went into high school and college, I finally gave up on trying to figure God out—it was just too much for me. In addition to the self-harm, I also began abusing different prescription pills and alcohol, and became very promiscuous.

Throughout middle school and high school, when I should have been connecting with people my own age, I stayed very isolated because I had a hard time trusting those around me. I just assumed that everybody would hurt me like my family and friends did when I was younger, so I decided I was not going to allow that to happen again.

In college, I met some people who were real and genuine. They seemed to truly believe what they said they believed. They would share about their relationship with God and it made me curious about who He really was. I was very drawn to the peace and joy that I saw in these girls, so I decided that I wanted my life to be different as well. As I began to take a look at the things that I was dealing with and the many problems that I had, I realized that I was going to need help in order to get out of this pit.

During this time, I heard about a program called Mercy Ministries. I was also struggling with an eating disorder at the time, and on the outside, that was the problem that was the most out-of-control in my life. I knew this was one of the many issues that they dealt with, so I decided to apply to the program.

When I first came into the program, I was afraid to share about my struggle with self-harm. I didn't want to give up control, and I had a difficult time trusting the staff. After I started feeling the genuineness of the unconditional love that was offered, I was able to begin opening my heart to God. However, because of my background with the occult, I was still scared of anything that had to do with religion and God.

I remember going to church and just crying the whole time. The staff truly cared about me and was always there for me, but I had a hard time communicating what was really going on in my heart. I was just crying and could not figure out why. I now know that God was beginning to soften my heart. I felt God's love as He was revealing His mercy and goodness to me. Even though I felt like I really didn't deserve it, God began to bring good things into my life.

I began to experience God's love more and more, especially through times of praise and worship. The comfort that would come over me was incredible, and I eventually came to understand that it was God's way of touching my life and letting me know He was there. The staff members were patient with me, and they took the time to explain to me what I was experiencing. As I grew and learned, they started teaching me how to go to God's Word and apply it to my life. They showed me how to turn to God in my time of need so that even after I graduated from the program, I would know how to continue to walk in freedom. I was amazed when I would open my Bible and read exactly what I needed to hear. I was in awe at how easy it was for God to speak to me through His Word.

An example of one of those times was when I opened my Bible to Colossians 1:21–23. This passage says, "This includes you who were once far away from God. You were His enemies, separated from Him by your evil thoughts and actions. Yet now He has reconciled you to Himself through the death of Christ in His physical body. As a result, He has brought you into His own presence, and you are holy and blameless as you stand before Him without a single fault. But you must continue to believe this truth and stand firmly in it. Don't drift away from the assurance you received when you heard the Good News. The Good News has been preached all over the world, and I, Paul, have been appointed as God's servant to proclaim it."

This passage was so detailed and specific to the lies I believed. Because of my involvement in the occult and because I had done so many wrong things, I did not believe that God could give me new thoughts and actions. I didn't know that it was possible for me to be able to think and live differently. However, this verse spoke directly to my heart and I was floored at the impact! Time and time again that kept happening and God became so real and personal in my life.

God was working in my life by softening my heart and helping me learn to trust my counselor at Mercy. I began to understand that my counselor was God's ambassador sent to me to help me work through my confusion. When I began to experience God's love and see that there was more to Him than I could ever imagine, I was ready to deal with the issues that had caused me to want to self-harm. Mercy provided an environment where I felt safe for the first time in my life. I started to open up and allow God to heal my broken heart. When that happened, my life was changed.

While I was at Mercy, I also had to deal with the anger I had in me. I was acting out one day and my counselor reached out to me and brought me into her office to talk about what was really behind all of the anger I was expressing. She prayed for me and explained that as Christians, we have authority through the name of Jesus over the spiritual forces that come against us. Through prayer, she came against the demonic forces and the anger that had such a grip on my life. I was still learning about what it meant to have authority in Christ, but I physically felt something happen that day. I knew that the demonic realm was real because I had seen scary things happen through the occult when I was younger. But, when my counselor was praying for me, I felt this relief. I literally felt the anger leave me—I felt lighter and could breathe so much easier. I was just like WOW—not only was the demonic real, but now I knew that God was real because I had felt His power. Not only did I learn that the power of God is real, but God's power is so much greater than the power of the enemy!

Another thing I had to face at Mercy was the abuse I had been desperately trying to forget. It was so hard to bring those painful memories before God and remember awful things that had happened. I didn't think I could do it, but God always carried me another step forward in my healing and replaced those painful places and emotions with joy, peace, and hope. I had to make the decision that I was not going to allow my past to determine my future!

I received healing and a deeper understanding of God during my time at Mercy. After several months, I reached a point where I was ready to start living on my own with the tools they had taught me in the program. It wasn't until after my time at Mercy that I had the opportunity to use these tools.

Only four months after I graduated, I had some really tough memories of abuse come up—I thought that I was going to crumble. Doubt tried to overwhelm me, but like I learned while I was in the program, I went to the scriptures and began reading out loud the promises of God for my life. God was so faithful to help me through those difficult times. Through talking to God and my accountability person about these memories and how they made me feel, I was able to get through it. It was yet another step in my healing process to realize that I am free from the abuse and that I am now safe.

Being connected to a healthy church is extremely important as I continue to walk out everything that I was taught at Mercy. While I was in the program, we attended church regularly, but now that I have graduated from the program and am living on my own, it is up to me to choose which church I go to. Because of the negative experiences in my past, I found myself going to the church we went to while I was at Mercy because it was safe and I felt comfortable there. I know that God is patient with me as I continue to take small steps and trust Him with every detail of my life, including where I go to church.

It really is all about *choice*. I found out while I was at Mercy that I do have a choice, and now I am free to choose the right thing because I am no longer controlled by the bondages of the past. Hurting myself whenever I feel guilt, hurt, or pain is no longer an option. If I have those feelings and emotions, I know I need to go to God and deal with the source of that pain. God continues to heal me more and more every day. I no longer even consider self-harm or any other addictive behavior, because it doesn't really resolve anything.

Cutting may provide a moment of relief, but it leads to a lifetime of pain and bondage. Now that I've tasted freedom and know how incredibly good God is, I never want to go back to my old ways of dealing with my hurt.

—Alyssa

Personal Study Guide—How does this apply to my life?

1. Alyssa believed that she was worthless and deserved to be punished. What do you believe about yourself that is leading you to self-harm?

____

____

____

2. God created you and sent His Son, Jesus, to give His life for you so that you can be free. This gives you incredible value and worth! What does Isaiah 43:1 say about who you really are?

3. Alyssa used self-harm as a way of escape. What emotions or experiences in your life are you trying to escape from?

4. God is patient and will wait for you to come to Him so He can heal you and set you free. In the Garden of Eden, Adam and Eve sinned and tried to hide from God out of a deep sense of shame. Read in Genesis 3:21 how God responded once they turned to Him instead of trying to escape. Even though there were consequences for their sin, what does this verse show you about the character of God and His grace?

5. In order to keep the abuse and self-harm a secret, Alyssa began to isolate herself from the people who were trying to help her. Identify secrets that you are holding onto in your life. What is keeping you from telling someone about those secrets?

6. Alyssa's healing process involved taking baby steps. What are some small steps you can take today to expose the secrets of your past so that you can find the healing and freedom you desire? For example, a baby step may be to avoid isolation, taking steps to find someone you can talk to, or even throwing away objects you use to harm yourself.

Talk to God

Use this space to write down any prayers, thoughts, or feelings you may have. This is a place to journal how you really feel.

Scripture to Study:

"But if we walk in the light, as He is in the light, we have fellowship with one another, and the blood of Jesus, His Son, purifies us from all sin."

—1 John 1:7

Chapter Two

BETHANY'S STORY: A PERFECT LIFE

There is no such thing as a perfect life. On the surface I seemed to have the "perfect" life—I had a great family who loved me and was very involved in church, but behind that image were dark secrets of sexual abuse and rape by a family friend. I didn't think this kind of stuff was supposed to happen to Christians, so I assumed something was wrong with me. I thought I had somehow brought it upon myself.

I had so much pain and guilt that I didn't know how to voice it. Even if I tried to talk about it, I was convinced no one would understand. I felt guilty for not being able to uphold the high expectations of others, and decided that cutting would be a secret way to punish myself.

It was never really a conscious decision to start hurting myself. I was just sitting around, dwelling on my pain when I started scratching my skin, and this quickly progressed to cutting. I just stumbled into it, but found some sort of release in the process. Even though it started innocently, it quickly became a very conscious decision that led to an addiction.

I experienced a lot of transition with school and church that was very difficult to deal with. After being home-schooled all my life, I started attending public school in 6th grade. That same year, our senior pastor and children's pastor left the small church where my parents were in leadership. They were people that I was very close to and really loved. The next year, I was home-schooled again. But when our youth pastor stepped down as well, my parents sent me back to public school again. For the next two years, there were constant changes in our church with different pastors coming and going. As a result, many families were also coming and going. I did not handle the changes very well, and I continued to use self-harm to deal with feeling so out of control.

My life became a facade—I felt like I had to pretend I was "fine" at all times and that I had everything together. Fearful of my issues being exposed—both the abuse and the cutting—I kept everyone at a distance. I went out of my way to make people angry with me so I would not have to deal with anyone getting "too close." I had a real fear of relationships. One of my friends discovered that I had been cutting myself. Not knowing how to help me, she pulled away and we were no longer friends. Her rejection confirmed everything I had been so afraid of and caused me to withdraw further into myself.

Since I had grown up in church, I knew all the "right" answers, but I couldn't reconcile a loving God with a God who had allowed me to experience so much pain. I saw Him as someone who wanted to punish me because I was a failure. I really believed that God couldn't love me because I was such a terrible person. I finally got to a point where I was sick of living a lie. I knew there had to be something more to life, and I was ready to do whatever it took to get help. I heard about Mercy Ministries while I was at college, and knew right away that it was exactly where I needed to be. I applied to the program and took a break from school when my application was accepted.

Coming to Mercy was not the easy way out of the mess I was in, but for me, it was the only way out. I realized that I could not hide the underlying causes of my behavior. I couldn't just focus on changing my behavior; I had to address the root issues. As long as I kept covering them up, nothing would ever change in my life.

The hardest thing for me was learning how to voice what I felt. Since I had expressed my pain through self-harm for so long, I had a hard time communicating what I was feeling. I also struggled with actually telling someone what was going on because I didn't feel I was worth their time or concern. Learning how to communicate came with having the staff at Mercy around me, who were very patient and understanding of the fact that it was a challenge for me. Slowly but surely, I learned how to verbalize the pain of the abuse and how I had kept it inside for so long.

I learned to trust and open myself up to the staff, and it eventually became easier. I was overwhelmed at God's goodness and how He is really concerned with every detail of my life. I came to know God personally and realized that I did not need to keep punishing myself because Jesus had already paid the ultimate price for my mistakes—He shed His blood and gave His life for me. Jesus died just for me because

I was worth it, not because of anything I had or had not done. It is not up to me to pay for my sins or for the sins of others because Jesus paid the price once for all.

I learned to believe that there were people who really wanted to know what was going on in my world and genuinely cared about how I was doing. God brought people who were willing to take the time to help me figure out how to put things into words. I had to learn how to recognize people who are safe and people who are not. It was a slow process of learning how to open up, and I had to ask God to bring people into my life that I could trust. I also had to be willing to take risks, not extreme ones, but little ones, and see how people would respond to me.

I still have my days where I don't know how to put my emotions into words. I found that writing is an easier way for me to communicate. I also played the piano for years, and that, for me, is another way I can express my emotions. When there are times that I do not know how to say what I am thinking, the creative expression I feel through playing the piano helps me express my feelings without using words. Sometimes just writing or drawing helps bring some clarity to what is going on in my head.

After I graduated from Mercy, it was challenging to go back into the same environment as a different person because people did not know how to respond to me. I had to trust God to restore the relationships in my family. At times, I questioned if I had really changed and began doubting whether or not I was different than when I had left for Mercy. I went back to what I learned at Mercy and responded to my own questions by asking myself what the lies were that I was tempted to act on so I could replace them with the truth. I reminded myself of the promise that God is going to finish the work that He started, and that I really am the new creation that He says I am. I went back to my old journals and read entries that reminded me of how far I had really come. Now, when I question the change, I have actual proof to remind me of the work that God has done in my life!

The biggest thing I have found in all of this is that I cannot expect the maturity process to take place overnight. I have to keep renewing my mind to what God's Word says about me. I know that as I am faithful with what I have been given, God will continue the good work He started in me (Philippians 1:6). I also had to set goals for myself. For example, after going a certain amount of days without cutting, then I knew that I could go another day. Then I could look back and say, "Wow! It has been months and I didn't even realize it!"

When failures and mistakes come, I cannot get stuck there—I have to learn from them and keep going. I realize that there are new levels of growth and new things to deal with every day. When I fall, I recognize what tripped me up so I won't fall again. I am learning not to discredit the ground that I have gained, but to also learn from my mistakes.

Harming myself was not a solution because it never fixed anything. Jesus took my punishment and pain on the cross, and nothing I do or fail to do will change that. His love is unconditional and the price He paid was sufficient to wash away my need to hurt myself. When I realized and accepted that, I was able to open up to Him and He was able to heal me.

When I realized that my relationship with God was based on His love for me, I could start being myself. No longer did I have to strive to be a perfect Christian because there is no such thing. I am able to do more now than I ever could have before, because I am at peace with myself, and my identity is in Christ. I am able to genuinely love the people He has placed in my world rather than hold them at a distance. That gives me incredible opportunities to be a blessing to others. Not only has God shown me my value, but He has also shown me how valuable others are, and He has helped me demonstrate that to them.

I just finished my third year at Hillsong College in Sydney, Australia and will continue to get my Bachelor's Degree in Theology. I love the idea of churches across the world that have different ministries that help their local communities, and I want to be a part of that!

—**Bethany**

Personal Study Guide—How does this apply to my life?

1. Bethany felt like she was expected to have it all together and could not be open and honest about her real emotions. In what ways do

you relate to Bethany? What specific emotions are the hardest for you to verbalize?

2. What do you think would happen if you were able to express those emotions to a safe person in your life? Identify any specific fears that you have.

3. When Bethany realized that she was unable to meet everyone's expectations, she began to cut to punish herself. How do you feel when you are not able to please everyone around you?

4. God's love and power are perfected through our weaknesses. Instead of focusing on the imperfections that you see in yourself and striving to meet impossible standards, rest in the grace that God gives you in this promise, because apart from Him, you can do nothing.

How does 2 Corinthians 12:9–10 release you from feeling like you have to strive for God's love and acceptance? What attitude should you have instead?

5. God desires a relationship with you simply because He loves you. His love is not conditional, nor is it based on what you do. God wants you to know that He loves you for who you are, and He wants you to receive His love without striving to earn it. Read Philippians 4:13 and write the areas where you need God's strength.

6. What steps can you take today to be honest about your true feelings and express the areas where you have experienced pain? Remember—His strength is perfected in weakness!

Talk to God

Use this space to write down any prayers, thoughts, or feelings you may have. This is a place to journal how you really feel.

____

____

____

____

____

____

____

____

____

____

Scripture to Study:

> "As for God, His way is perfect; the word of the Lord is flawless. He is a shield for all who take refuge in Him. For who is God besides the Lord? And who is the Rock except our God? It is God who arms me with strength and makes my way perfect. He makes my feet like the feet of a deer; He enables me to stand on the heights. He trains my hands for battle; my arms can bend a bow of bronze. You give me your shield of victory; you stoop down to make me great. You broaden the path beneath me, so that my ankles do not turn."
>
> —2 Samuel 22:31–37

Chapter Three

HEATHER'S STORY: THE MONSTER IN THE MIRROR

When I was five years old, my newborn baby brother died. This was obviously a very difficult time for my parents, and this contributed to my childhood being very dysfunctional. I lived in constant fear and chaos, wondering what would happen next. My hopes of having a "normal" family were shattered.

Furthermore, my father became an alcoholic and abusive. As a result, the fear and shame intensified. I felt worthless and thought everyone was out to find the worst in me. As a teenager, I would wake up in the morning consumed by fear and worry. After a normal conversation with just about anyone, I would walk away questioning whether I said something in the wrong tone, the wrong way, or if they were judging me. I was afraid everything I did or said was wrong. I feared that people were going to leave me and that I would be all alone. I dealt with these fears by attempting to control what I ate, hoping to "look perfect." This is where the eating disorder began.

My family did not know how to respond to me so I withdrew from everyone, feeling rejected and misunderstood. I did not want to be around anyone because I was controlled by so much fear. My family's response did not make the situation any better. When they would see me during the holidays, they would make rude comments and question me about my low weight, which had become very noticeable.

Based on the history of mental illness and addiction in my family, I thought I was genetically destined to live a life of despair. I felt certain my existence was a mistake because everything I did or said was wrong and unacceptable. I believed I had to punish myself for being alive, so I began to cut. I used cutting as a way to remind myself not to mess up, speak, act, or even eat. I felt hopeless, depressed, and distraught. When I felt full of these negative emotions, physical pain seemed to take away the mental anguish, at least temporarily. Cutting distracted my mind from focusing on internal pain, displacing it to the external.

It tore my mother apart to know that I was cutting. Imagine seeing your daughter with the word FAT cut into her arm. My friends started to withdraw from me because they did not understand how I could do that to myself. They were Christians and knew that I was too, so they couldn't understand why I saw a monster in the mirror.

I felt that God was far away from me. I always knew in my head that He was there for me, but I did not feel it in my heart. I longed to be close to Him, but I could not figure out a practical way to get there. I could not understand why I was going through so much pain, and why I wasn't receiving any relief from the anguish. I had so much anger towards God, myself, and my life that I shut myself off emotionally.

Depressed, suicidal, and sick with anorexia, I couldn't bear to live another day. Paralyzed by fear and totally unable to function, I finally reached the end of my rope. I *knew* abundant life was possible because I had seen others walk in freedom, but I had to choose it for myself. I heard about Mercy Ministries from a friend, and applied to the program, truly desperate for help. I entered Mercy a few months later and started the process that began to change my life.

One of the hardest things for me was getting out of that "victim" mentality. I honestly believed I would not be loved unless I was sick, and that my dysfunctional behaviors would earn me the love and attention that I longed for. I had to learn to accept the love that my Heavenly Father gives me, knowing that only His love is perfect and casts out all fear (see 1 John 4:18). As I began to speak the truth of what God's Word says about me out loud, my life began to change. Learning to speak the Word of God was not something that I was taught to do in any of my past treatment centers, but was definitely a huge key to my freedom. Faith came as I listened to and spoke God's Word over and over.

I also learned that I am the righteousness of God. Righteousness means to be made upright, worthy, and free from guilt! God focuses and dwells upon what is right with us, not what is wrong with us. He died so that I may live with acceptance, and be in a right relationship with Him. I learned how to use the truth in God's Word to overcome the lies fed to me by the devil.

Also, the prayers of Godly people surrounding me were crucial. At Mercy, I was always able to pray with someone whenever I needed to—something that was never offered at any of my past treatment centers. The staff always made themselves available to me, and I began to understand God's love for me through the unconditional love that I was shown at Mercy.

I learned about the generational patterns that ran in my family and realized that I am no longer doomed to the sins and addictions that I was born into. There is a huge history of alcoholism, divorce, gambling, depression, suicide, sickness, and a variety of dysfunction in my family. Before I came to Mercy, I didn't know what a generational pattern was, but as I started to look at my grandparents, aunts, uncles, and other extended family members and the issues they struggled with, I was able to identify certain things. I learned that I have the power to choose to forgive everyone, including my own family members, and make different choices to prevent these family patterns from continuing. Forgiveness was another key—I had to choose to forgive them and then I had to forgive myself for falling into the same patterns.

During my time at Mercy, the staff helped me address the issues that were behind the fears that had consumed my life. I wrote down all of the fears that were in my head. Things such as, "if I say something wrong, this person is going to stop loving me." I wrote out my irrational thoughts, fears, and insecurities on a piece of paper. By the time I was done, I had written out over 15 pages of tormenting thoughts, which were actually ungodly beliefs. I confessed that I had believed those lies that the enemy had placed in my mind. I went through my list and found a scripture to counter each lie. I spent time personalizing the scriptures and speaking them out loud until the truth became more real and powerful than the lies I had previously believed.

I learned that God is for me and not against me. I realized that I had been closing the door on God and His power to heal. I learned that He fulfills His word and His promises to heal and restore. Jesus died on the cross to disarm all of the rulers and authorities, triumphing over them, and He has given ME the power to do the same in His name! I learned that I do not have to lie down and let the enemy defeat me, but that He has given ME power to defeat Satan! The name of Jesus is above every name, including self-harm, self-punishment, abuse, hatred, anorexia, and fear. I am made in the image of God, and my body is not my own. Therefore, every time I hurt my body, I was hurting God. I believe He cried every time I chose to hurt myself.

In addition to dealing with the root issues, I also had to deal with the physical temptation to harm myself. I overcame the desire to self-harm by realizing that my body is not my own. I have been bought with a high price and my body is the temple of the Holy Spirit (see 1 Corinthians 6:19, 20). My responsibility is to care for my body in the way God cares for me and to choose victory daily. Now, I am so passionate about making decisions to choose life! The Bible showed me that God does not force His will upon us. He said it Himself in Deuteronomy 30:19: "I have set before you life and death . . . now choose life." Choosing life is a daily decision to live and to choose God's way.

There are times when I am dealing with negative feelings and emotions, and the old thoughts of harming myself to stop the pain creep in. My desire to self-harm is also triggered by stress and fatigue. If I don't get enough sleep, I am not able think clearly and I don't have my guard up. If I have not been doing what I need to be doing to keep my relationship with the Lord strong (reading my Bible, going to church, praying, and sharing openly with an accountability person), I slip into old ways of thinking. In those moments, I have the opportunity to use the tools I was given during my stay in the program at Mercy.

Instead of giving in to self-harm, I usually go for a walk or ride my horse. Listening to worship music really helps, as well as taking a hot bath. It's important to wait for the really strong emotions to pass, and if someone I trust is available, I will go talk to them and ask for prayer. Once the emotions pass, I can look back and see what triggered my desire to cut, and how I can prevent it from happening again. It is very important to identify the root causes of what triggers this desire. Once I do this, I can find scriptures that apply to whatever areas are identified and combat the lies with God's truth.

I know that when God wrote out the plans for my life, He did NOT include self-harm—neither did He include the negative emotions and situations that led me to cut in the first place. His plan is to prosper me and to see me living in victory! It breaks His heart to see me in this kind of pain. Every tear I cry, I believe He cries along with me. He is waiting with His arms outstretched for me to run to Him and fall into them safely. What He has done for me, He can do for you!

For the first time in my whole life I can say, "I love me!" I know that I am a beautiful, precious daughter of the King! I am magnificently created by the Creator of the universe! In Him, I have everything I need—all of my needs in every area are met. I see myself as a victorious, powerful, woman of God with an amazing future ahead!

At the time I was engaging in self-harm, I never thought that I would have to look at those physical scars everyday for the rest of my life. When I first graduated from Mercy, the scars really bothered me and I was embarrassed and ashamed for anyone to see them. Now, they remind me of what a good God I serve. My God has delivered me from so much pain and torment. I am living proof that God *can* and *does* restore all that was lost!

—Heather

Personal Study Guide—How does this apply to my life?

1. After being sexually abused, Heather felt full of shame which led her to self-harm. What past experiences in your life have led you to feel shame?

____

____

____

____

____

2. John 10:10 says that Satan has come to steal, kill, and destroy, but Jesus comes to bring life. As you read Jeremiah 29:11, what else does God show you about His plans for you?

____

____

____

____

____

3. Despite what Heather's friends told her, all she saw was a monster when she looked in the mirror. What do you see when you look in the mirror?

____

____

____

____

____

4. The truth is that you were created in the image of God (Genesis 1:27). Read Psalm 139 out loud and write what this passage reveals to you about your true value to God.

____

____

____

____

____

5. Heather's sense of shame eventually led to fears that she would not be loved or gain the attention that she longed for if she did not self-harm. Make a list of the fears you recognize in your own life that are keeping you in the cycle of self-harm.

6. 1 John 4:18 says that "perfect love casts out all fear." With the help of a mature Christian friend or counselor, make a list of truths from God's Word that contrast your previous list of fears. Throw your old list away and begin to speak the truth over your life!

Talk to God

Use this space to write down any prayers, thoughts, or feelings you may have. This is a place to journal how you really feel.

____

____

____

____

____

____

____

____

____

____

Scripture to Study:

> "Because the Sovereign LORD helps me, I will not be disgraced. Therefore have I set my face like flint, and I know I will not be put to shame."
>
> —Isaiah 50:7

Chapter Four

JESSICA'S STORY: CONSUMED WITH LIES

When I was five, I became an actress. I pretended to be what I was not to escape the reality of being sexually abused. I kept silent about the abuse, believing my life would be in danger if I told the truth. My mind was consumed with lies that I was dirty, ugly, and that the abuse was entirely my fault.

The lies I believed were so numerous that I considered my whole life to be a lie. These were not lies that a child would normally tell a parent, but lies to cover up who I really was. I had a skewed perception of everyone. If I didn't do everything I could to please the people around me, I believed I would be rejected. At the same time, I felt as though everyone used me and took advantage of me.

The struggle in my mind was so intense at times, that I questioned what was real and what wasn't. I soon found that the only way (so I thought) I could snap back into reality, was through cutting and beating myself. This pattern continued for many years and worsened as time went on.

God did not become a part of my life until I was twelve. My understanding of God was very distorted. He seemed to be the same as everyone else—hard to please and a user. When I was in the midst of harming myself, I had very few thoughts of God. It seemed as if He had disappeared and left me in my despair.

I thought if I just kept pretending, I could make everyone happy with me, so I tried to act like I was the best Christian and the best partier at the same time. Trying to live in both worlds, I felt like a super-hypocrite. I found that when I would hurt myself and it was noticeable, I had to lie to cover it up. Constant lying built a wall between me and the other people in my life.

However, most of the time, the other person I was lying to knew the truth about what was going on and was waiting for me to be honest about it. I noticed how selfish I had become, and was usually just concerned about me. This caused me to lose sight of what was truly important: God, family, and friends.

Life continued in a downward spiral as I was lying, drinking, cutting, and beating myself more frequently. I began to hurt my friends. One night, after making a comment I instantly regretted, I broke down and told my friend that I desperately wanted to change. I knew I needed some major help. My friend and I got down on our knees and prayed that God would help me find a place to go. We knew I needed a place where I didn't have to pretend to be someone I wasn't, or worry about what anyone else thought. The next day I got on the internet and God led me straight to the Mercy Ministries web site. After going through the application process and being put on the waiting list, I eventually entered the Mercy Ministries program.

At Mercy, the first thing they helped me do, was recognize the lies I had believed my whole life. I realized a lot of what I had previously believed to be true was actually deceptive. As I recognized this, I was able to replace the lies I had believed with truth. Most of the truths came from the Bible, and others came as I faced the actual facts of my family history rather than my fantasized version of reality. One of the biggest truths I learned, was that I am precious in God's sight. As a young woman I never felt worthy of anything, but now I have discovered that God has a plan, purpose, and a future for my life.

I understood for the first time that God did not want me to just serve Him, but to worship Him. To me, that meant I could have a relationship with Jesus Christ without feeling obligated. I do things for God now because I want to, not because I have to. I recommitted my life to Christ by inviting Him into my heart and life, and asking Him to forgive me of all my sins. I also made the choice to forgive myself and those who had hurt me.

Letting go of all the different masks I wore to please people (and even God), took a lot of humility and hard work. I had to get to the bottom of all the different faces I was putting on. I went from pretending to be who everyone else wanted me to be, to loving the person that God created me to be. I had to get to know myself.

It was a very slow process because it needed to be. Change takes time, and if God showed us everything we needed to deal with all at once, we wouldn't be able to handle it. It may have started with something little like, "my favorite color is green," but if my favorite color is really purple, then that was a lie and I really wanted to be like someone else. Over time, I was able to address the bigger lies of "I am not beautiful because of the sexual abuse or the self-harm." I had to start small and work my way through all of the deceptions that had become my identity, which in reality was a false identity. It took a while to work through the denial, but God is gentle and was so patient walking through the whole process with me. After graduating from Mercy, I had to discipline myself to keep instilling the truth in my heart and mind.

To replace the lies, first I had to identify them. Something I found helpful was writing them down. Some of these lies were that I was ugly, or that no one would ever love me because I had been abused, taken drugs, and had sex. After I wrote those things down, I wrote the truth to replace them. An example of this is: I am a child of God and He has a future and a plan for my life despite the mistakes that I have made. I also learned about the power of my spoken words. I had to speak the things that were true, even if I did not believe them yet, because as I speak the truth, faith rises in my heart to believe it. The spoken Word began changing me on the inside.

Another thing I had was a very distorted perception of forgiveness. I learned that forgiveness is a choice, not a feeling or emotion. Forgiving someone does not mean I am saying what they did to me was okay. However, it is necessary for me to forgive in order to receive God's forgiveness in my own life, and to experience internal freedom.

One person that I had a hard time forgiving was my dad. I had a lot of anger toward him because he never trusted or believed anything I said. I experienced a lot of verbal abuse from him. Even when I returned home after Mercy, it was a continuous process of walking in the forgiveness that I had chosen to extend to him. I saw that as I walked in forgiveness, my emotions began to stabilize and I was able to love him more and more. Over time, I felt like I was able to show him true Christ-like love.

My dad recently passed away after suffering from a severe illness. Because of the example I was able to show him through loving him and forgiving him, He accepted Christ the week before He died. It was so painful to watch him go, but it was the most beautiful thing that I had ever seen because I knew that he truly died in peace. If I would have still had unforgiveness in my heart, I would have been unable to come home after Mercy and love him in a new way. It's still a process. I still have to choose to forgive at times because my heart can grow cold in some areas. I am still human and think about things from the past, but I have to keep pressing forward and continue seeking God in those tough moments.

Mercy did not just deal with my symptoms, but went down to the root issues in my life. Mercy also gave me the tools I needed to find freedom and keep it after I graduated the program. Staying in fellowship with other believers, finding a solid Bible-teaching church, and reading God's Word daily, are some of the keys I have to use to walk in freedom. Daily I am digging into the Word, praying for God's wisdom, grace, love, and revelation, communicating with amazing Godly people, and making sure that I am connected with my church. All of these things have allowed me to stay free from the bondages I was once trapped in.

Now I see myself as a beautiful woman of God, walking in Christ (and Him in me), and seeking all He has for me. If I truly want to get to know God and be more like Him, then I have to spend time with Him. Just like if I want to be best friends with someone, I have to spend time with them to learn who they really are and know them for myself, even if I have heard other people tell me about them. I have learned that things like reading the Word, going to a good Bible-based church, and finding true fellowship with other believers are really important. Find people that are willing to walk with you through the hard times, who love you for who you are, and encourage you to seek God with all of your heart.

It is hard to believe I have come so far, but I give all the glory to God for how He has restored me. Now I am willing to travel and share my testimony with whoever needs to hear it. I also have a desire to bring a healing touch to other young women who are in need by eventually working for a ministry such as Mercy. I know that whatever God has planned for me will be amazing and exciting!

I praise God for setting me free from all the things that bound me, but it was a struggle to get here. I found freedom when I stopped focusing on myself and started looking upward to a faithful and loving God who desires to see me walk in freedom.

—Jessica

Personal Study Guide—How does this apply to my life?

1. Jessica believed that she was dirty, ugly, and that the sexual abuse she experienced was her fault. What similar lies do you recognize in your own life?

____

____

____

____

____

2. John 8:44 says that Satan is the 'father of lies.' How does knowing that Satan is the source of the lies you have just listed above make you feel?

____

____

____

____

____

3. When you continue to speak a lie, you eventually believe it as truth. Can you remember a time where you had a difficult time distinguishing truth from deception?

____

You may not yet recognize areas where you are being deceived because deception is just that—the inability to recognize truth. Take time to pray and ask God to reveal the areas of your life where you are living blinded by deception that you may not be aware of.

4. The lies in Jessica's mind became her distorted perception of truth, leading her to other destructive behaviors. How have you seen the same progression in your own life where you have believed a lie and your actions reflected your belief?

5. To eliminate the lies behind the behavior, you have to bring truth to the root of it. What truths are in Psalm 16 that you have had a hard time believing could be true for you?

   Take time to go over the things you identify with a trusted friend or counselor, so someone can help you process why you have a hard time accepting those truths.

___

6. Reading God's Word daily is vital in order to grow in a relationship with Him and to renew your mind to the truth. What steps can you take today to begin renewing your mind to God's Word? An idea may be to write out Bible verses on index cards and read them out loud, saying about yourself what God's Word says about you. Set aside a time each day to read the Bible so you become more familiar with truth.

___

___

___

___

___

___

Talk to God

Use this space to write down any prayers, thoughts, or feelings you may have. This is a place to journal how you really feel.

Scripture to Study:

"You are my lamp, Oh Lord. The Lord turns my darkness into light."

—2 Samuel 22:29

Chapter Five

KATIE'S STORY: A SILENT STATISTIC

At four years of age, I began to "discreetly" punish myself. I believed I was "bad" and began wounding, depriving, and overworking myself because I thought I deserved to feel pain. As I grew older, my self-inflicted wounds became more serious, often sending me to a doctor or the emergency room. I never really thought about why I intentionally hurt myself, but I convinced myself that I did not deserve to feel good. I didn't think I deserved to be alive either.

I didn't understand that, with each injury, I was taking another step down a path that would only lead to further spiritual and emotional death. The choices I made to hurt myself were feeding right into the enemy's plan for my destruction. I was in bondage and had no hope for healing of my deep internal wounds.

During my entire childhood, I endured severe physical, emotional and sexual abuse, as well as neglect, at the hands of several different individuals. I had begun to repeat the cycle by abusing myself, though I was not fully aware of what I was doing. As my depression became more and more severe, I believed that hurting myself was my way of coping, my way of living.

Even though a church-going family raised me for fourteen years, the God they modeled to me was harsh, demeaning, and cruel. My parents were extremely religious, but also extremely abusive. I was deceived into thinking that God approved of the things that they would do to me. My father would use the verse from the Bible that says, "Children obey your parents in the Lord for this is right" (Eph. 6:1). He told me that this verse meant that he owned me and that he could do anything he wanted to me. He said that I had no right to stop him because this was a divine right that God had given him. He had completely distorted the scripture and taken this verse out of context, using it to control and manipulate me. I believed that I must be bad and I deserved the abuse. I felt that punishing myself was the right thing to do.

My life as a young teenager was in a constant state of disarray. I was continually tossed back and forth between foster homes, mental hospitals, and other state facilities. I clearly remember one specific day, sitting in the emergency room after being taken there by force in an ambulance. None of the hospital staff knew what to do with me. The psychiatric hospital I had been released from only four days earlier, turned me away because they no longer believed they could help me either. The intervention staff in the emergency room decided to send me to long-term care in the state psychiatric ward—the kind of hospital many people never leave. I felt utterly hopeless when the staff informed me that an ambulance would be arriving shortly to take me away. I cried and begged them for one more chance. I wanted them to believe I could be something more than another silent statistic that faded into the system as just a number that was soon forgotten.

By the grace of God, they sent me to a different hospital and I was not committed to long-term care that day. I was thankful, but I knew that I needed a miracle or I was going to die.

During my various placements, I had heard about Mercy Ministries and had always felt drawn to it. While I didn't understand why or how, I felt certain that I needed to be there. I wanted so badly to be healed and made whole, but I didn't really believe it was possible. Yet somehow I knew going to Mercy was my only hope.

I eventually went to Mercy Ministries and God began to do an amazing work in me. My time at Mercy was unlike anything I had ever experienced. I was introduced to Jesus, and He became my Savior, my LORD, and my friend. I came to know the God of the universe who uniquely designed me exactly as I am for a specific purpose (Psalm 139). I learned that He is my best friend (Isaiah 41:8), my Refuge (Psalm 90:1), my Father (Psalm 68:5), my Healer (Isaiah 58:8), and my soul's resting place (Psalm 62:5). I met my Redeemer, who took on the punishment for all of my sins (Isaiah 53:5), and who truly hurt every time I injured myself (Ephesians 4:30). I learned that God would not waste my life, but would take it and transform it into something beautiful.

Mercy was a house of healing for me. There, God showed me that in the midst of everything I had endured, He had been protecting me. He had never left my side and He had a purpose for my life, even beyond the pain. While I was at Mercy, I learned that God did not withhold good things from me. He had preserved my life for a good reason. I now know to trust and love the only one who can fulfill me, who longs for my company, who cherishes me, and who knows every part of me. He has looked into my heart, declared me clean and pure (Titus 2:14), told me to never forget that I belong to Him (1Thessalonians 5:5), and to never call myself dirty again because He has made me clean (Acts 11:9).

When I am struggling, I talk to someone I can trust. I have to lay down my pride and tell them when I am having a hard time. I am doing great in so many ways, but sometimes it is still hard for me and I have to recognize old triggers. God has placed some incredible people in my life who are willing to really listen to me and who help me identify and deal with triggers in my life, or in some cases, avoid what triggers me altogether.

Another important thing for me was learning to cry! I did not let myself cry for years. I have found that crying is such a great way for me to express my emotions and release the grief and pain of my past. It still feels unnatural at times because I suppressed my feelings for so long, but there is such a release in just letting myself cry.

I used to believe self-injury was my only option, but I now understand that self-harm is a choice and Christ has given me a way out! He took my shame and my issues, and bore the burden of them. Isaiah 53:4 says, "Yet, it was our weaknesses He carried; it was our sorrows that weighed Him down." To think, we thought His troubles were a punishment from God, a punishment for His own sins! He provided the path for my freedom. His blood is sufficient to pay the price for all sin (1 John 1:7), including the sins others have committed against me and sins I have committed against God and others. I now know the comfort and peace of God and I have felt the embrace of my good, gentle, and perfect Father.

Since my graduation from Mercy, God has given me opportunities to help others who deal with the same issues I used to have. He has even allowed me to play an integral part in their healing. The changes in my life are amazing, and I thank God for allowing me to go to Mercy to give me another chance at life and to receive freedom. My greatest desire is to live the abundant life that Jesus died to give me, choosing daily to surrender to him and not to the pain of my past.

God has touched my heart and changed me forever, so I no longer desire self-injury. He has drastically turned my life around. Before, my life was full of pain and hopelessness, but today I am pursuing a medical career in which I will be able to help bring physical healing to others!

—Katie

Personal Study Guide—How does this apply to my life?

1. Katie was convinced that she deserved pain and death, believing that she was bad. What do you believe that you deserve?

____

____

____

____

____

2. When did you begin to believe that about yourself? Write down the specifics of that memory and talk about it with a mentor or mature Christian. Be sure to identify where that false belief began in your life.

____

____

____

____

____

3. The God that was portrayed to Katie through her parents was harsh, demeaning, cruel, and the cause of her pain. How is this contradictory to the truth stated in Psalm 86:15 about who God *really* is?

____

____

____

____

____

4. Katie begged God for another chance, fearing that she was forgotten. Identify a time that you have felt forgotten and alone.

   The truth is that you do not have to beg God, because He promises He will never forget you (Isaiah 49:14–16). What does Isaiah 59:1 say about God's ability to save you when you turn away from your sin, regardless of how hopeless you feel?

____

____

____

____

____

5. Even when you see no hope in your circumstances you can always find hope in God. What does Isaiah 40:31 say will happen when you choose to put your hope in the Lord?

____

____

____

____

____

6. Katie identified things that triggered her to cut so she could take preventive measures. What things trigger your urge to cut or self-harm and what can you do to avoid these triggers in your life? Share these things with someone you trust so they can help hold you accountable.

____

____

____

____

____

____

Talk to God

Use this space to write down any prayers, thoughts, or feelings you may have. This is a place to journal how you really feel.

____

____

____

____

____

____

____

____

____

____

Scripture to Study:

> "For I know the plans I have for you, declares the LORD, plans to prosper you and not to harm you, plans to give you hope and a future."
>
> —Jeremiah 29:11

Chapter Six

NATALIE'S STORY: I MADE A CHOICE

As the baby of my family, I was expected to learn from all the mistakes my older siblings made. My brother and sister were really wild growing up, and while I was still in elementary school, I saw them getting drunk and partying all the time. I also saw how much it hurt my parents when they got into trouble. My parents had always told me that I was the only one they could count on to be responsible and as a result I felt the pressure to maintain that "perfect child" role.

The perfect image I was trying so hard to protect was shattered after a family friend began sexually abusing me. From this experience, I developed deep roots of insecurity, looking everywhere for acceptance and happiness. I became a "people pleaser," doing whatever it took to make other people like me. The more I tried to please others, the deeper I fell into a state of depression and self-hate.

During seasons of transition and change in my life, I would cling to anything to maintain a sense of control. My brother had just moved away to Florida and my sister left for college, leaving me as the only child still at home. I began to self-harm as well as develop very abnormal eating habits. Before I knew it, I was addicted to self-harm and the "bad eating habits" turned into a life-consuming eating disorder. I knew that I was in trouble, but wasn't sure who to talk to about what was going on.

A new basketball coach came to my school and I began to open up to her about what was going on in my life. My relationship with her caused problems because the other girls on the team thought that I was her favorite. My teammates became very jealous because I was a starter on the team, even though I was younger than they were. They started rumors about my relationship with my coach that were simply untrue.

Their attacks just caused me to draw closer to her because everyone else on my team started to hate me. In an attempt to protect me from being ridiculed by my teammates, my coach started being verbally abusive to me at practice so the girls wouldn't think she favored me. If there was a bad play on the court, I got blamed. Everything became my fault. One time she was so angry with me after a game that she threw me up against the wall. I didn't think anything was wrong with how she was treating me because I still saw the actions as her being protective of me.

My parents did not approve of my relationship with her. They thought it was odd that we were spending so much time together. I disagreed because she was the only person that I was able to trust and open up to. They asked me not to have anymore contact with her outside of school. When my parents found out that I had gone over to her house, they forced me to quit the basketball team. I was devastated, and I completely shut down, refusing to talk to anyone. I began to self-harm more than ever, and I focused all of my energy on the eating disorder.

Mid-way through my senior year, my youth pastor sat me down to talk after he noticed that I had lost weight and was falling deep into depression. He did not know about the self-harm, but told me that he was going to talk with my dad and that he refused to stand back and watch me kill myself. I begged him not to, but he went anyway and had a talk with my dad. During my senior year, my parents finally sent me to a counselor. After going a few times I declared that I was fine and had no more problems. Everyone believed me, but the scariest part is that I actually believed it myself.

The next semester I went to college and my life completely spiraled out of control. I knew the eating disorder had started to get out of hand, but for some reason I didn't think the self-harm was a big deal. I still thought that I could stop myself from doing it if I wanted to badly enough. With all of the focus on the eating disorder, the self-harm stayed my secret. It was not until my doctor discovered the truth, that I began to realize cutting might actually be a problem. I had never felt as ashamed as I did that day at the doctor's office, but instead of stopping, the shame drove my desire to cut even more.

I kept razors in my purse, pockets, and everywhere else just for the comfort of having them there. Each time I cut myself, I had a growing desire to do it more often and deeper. I had shut off all emotions. Even if I wanted to cry, there was nothing there, but somehow the pain reminded me I was still alive. Everything that happened around me fed the desire to hurt myself.

The following summer I was put in the hospital for the third time, and my doctor told me that I couldn't leave until I had a place to go. My parents were calling other treatment centers, but money was quickly running out. I called my athletic director and told him that I was in the hospital and they were not going to let me out. He told me about a place called Mercy Ministries. I began the application process and before long, I was packing my bags to enter the program.

When I walked through the doors at Mercy, I knew immediately that something was different. God was in that house and I felt a peace that scared me. Everything in me wanted to run. I don't know if I was more afraid of getting better or of people reaching out to me, but I knew something about this place was different.

As much as I knew something had to change, the fear of letting go of my issues drove me to act out more. Despite my bad attitude, complete lack of trust, and total denial, the Mercy staff handled me with love and acceptance every day. I began to receive revelation after revelation of God's love for me through the staff, and even through other girls that were in the program.

One Friday night, we watched the movie *The Passion of the Christ*. As I watched the scene where people began to whip Christ, the realization hit me that every time I cut myself, I was punishing myself instead of accepting the fact that Jesus took my punishment for me on the cross. I was denying what He had already done for me. "Surely He took up our infirmities and carried our sorrows . . . but He was pierced for our transgressions, He was crushed for our iniquities; the punishment that brought us peace was upon Him, and by His wounds we are healed" (Isaiah 53:4, 5). This did not mean that I was expected to be perfect, but I had to make a conscious decision to accept the fact that I did not have to punish myself and accept the fulfilled life He died to give me.

Since graduating from Mercy, I have made bad choices that I regret, but I have also realized that is exactly what they are—*choices*. The times I struggle the most are times when I have stopped seeking God first in everything I do. I have learned I cannot be alone in my struggle; I need to surround myself with support and put the Word of God in my heart. I have also learned that I cannot keep certain objects in my room that I used to harm myself with before, because while those things may not bother other people, they are a temptation to me.

When I had only been out of Mercy for five or six months, I started having nightmares and flashbacks from the abuse that happened in my past. I called my former counselor at Mercy because I was so discouraged. I thought I had dealt with everything while I was in the program and didn't understand why these memories were coming up. She told me that there were still some areas that God wanted to heal in my life. She reminded me that God was still at work in my life, and encouraged me to allow Him to heal those painful places.

I had to swallow a lot of pride because I thought that after I graduated from Mercy, everything was supposed to be fixed. But I knew that this issue had to be dealt with, and I also knew that I had what I needed to get back on track. Slowly but surely, I began to deal with some of those painful memories.

To be honest, my first thought was to grab a razor. I found one and I just held it in my hand. I started shaking and I knew I couldn't do it. I knew that I was making a choice, and if I gave in once, that I would be pulled back down easily. I put the razor down and called someone that I trusted. I needed to let someone know what was going on so I wasn't trying to fight this alone. I got in my car and went for a drive, which got me out of my room and away from the temptation. It may not have been the best to go for a drive because I was really upset, but I put on some praise and worship music and just drove until I could think rationally.

There have been days and nights when there was just a lot going on in my mind, and my first thought was to go back to the eating disorder or the cutting. I have not been perfect, but I have learned that I have to tell people when I am struggling. I know that I can go to God in my time of struggle and that just because I have a thought to hurt myself, it does not mean that I have to give in. I have to take the thought captive and apply the truth of God's Word to the lies of the enemy. I know that I have made it through tough times before without having to cut, so I know I can make it again. If you can get through that first 30–60 minutes where the urge to cut is the strongest, you will feel so much better than the times you actually gave in to the struggle.

I now know I have a choice. I now know the power described in 2 Corinthians 12:9, "His strength is made perfect in my weakness." God has not given up on me and I know that He never will.

—Natalie

Personal Study Guide—How does this apply to my life?

1. Natalie did not choose the circumstances of her past, but she did have a choice in how she dealt with the things that had happened to her. How are you choosing to deal with the negative things you have experienced in your own past?

____

____

____

____

____

2. After making one bad choice to cut, it led Natalie to desire to do it again, becoming an addiction in her life. How has one bad choice led to another in your life?

____

____

____

____

____

3. God has given each of us the gift of free will, which means that you have a choice in how you deal with each situation in your life. What does Deuteronomy 30:19 tell you about the choices you make and what they ultimately lead to?

____

____

____

____

____

4. When God promises to bring you life, it means so much more than simply being alive and surviving from one day to the next. Read John 10:10. It is very clear from this verse that the enemy's plan for our lives is to steal, kill, and destroy. However, what does Jesus tell you in this verse about the kind of life that He wants you to live?

____

____

____

____

____

5. Even when the environment around you doesn't change, your perception of your environment can change everything. What does God promise in Proverbs 8:32–36 to bring you when you are obedient to His Word?

____

____

____

____

____

6. What choices can you make to choose life today?

Talk to God

Use this space to write down any prayers, thoughts, or feelings you may have. This is a place to journal how you really feel.

Scripture to Study:

"There is therefore now no condemnation to those who are in Christ Jesus, who do not walk according to the flesh, but according to the Spirit. For the law of the Spirit of life in Christ Jesus has made me free from the law of sin and death."

—Romans 8:1–2 (NKJV)

Chapter Seven

TIFFANY'S STORY: A WAY OF ESCAPE

Straight up: I *hated* emotions. I hated feeling angry, sad, hurt, or any other emotion that didn't make me feel good. I thought I was supposed to be happy all the time, and even when I went through hard times, I thought I was supposed to look like I had it all together. Because of those unrealistic expectations from others and myself, self-harm became a "quick fix" for me. I could be "happy" all day at family gatherings, school, and even church, but when I was alone, I had to face the true emotions I was really dealing with and get a release from the pressure to be perfect.

One of the biggest contributing factors leading me to self-harm was that two family members were sexually abusing me, and I believed I had done something to deserve this. I even remember thinking, "maybe when my abusers see the scars, they will be grossed out and leave me alone." I thought if I made myself unappealing to them, they would stop. I came to realize how much of a release cutting was for me and how good it felt. Cutting provided a temporary distraction from the inner pain and all the fear and insecurity inside of me. It was a way of escape.

My perception of God was that He was a distant law-maker. I thought I had to follow the rules, and I had never experienced a personal relationship with Him. The church that I grew up in was all about what you can and cannot do—I do not remember seeing a display of love or grace.

Even though I believed there was a "God," He wasn't a personal God to me. I didn't believe He cared about me or the things I did, because I wasn't perfect and I didn't feel I deserved His love. I was so angry and looked for someone to blame, so I blamed God for my pain when in reality, He was the only one who could truly free me from pain.

My dad was very physically abusive, and my mom was verbally abusive. I struggled daily wondering, "Did my parents love me for who I was? Did they really accept me?" They would tell me I was fat and that I would never amount to anything. I was also told that I was a mistake. After hearing those things, I believed that if my own parents didn't want me around, why would anyone else?

It was all too much and one night I planned to commit suicide. I didn't really want to die, but I didn't want to live this way either. Desperate for help, I called a lady from my church as a last cry for someone to care. She picked me up from my house and brought me to hers so we could talk. She was very compassionate and listened to me as I expressed all of my frustrations and the tangled mess of emotions that I could not sort out on my own. When I was done, she encouraged me by sharing that God really did have a plan for my life and that He was going to take all that the enemy meant for my destruction and turn it around for my good. She also told me that one day, I was going to be able to encourage other girls who would be struggling with the same things because I would understand them. Then, she told me I had to allow God to work in my life first. That hit me. I realized at that moment that I was being selfish and couldn't see past my own pain.

At that moment, I knew that things had to change in my life—I wanted to stop hurting myself, but I didn't know how to practically do that. I could go to church every Sunday and Wednesday and read my Bible, but I didn't know how to apply what I was reading to my life so that something could really change. No one had taught me what to do when I was struggling or how to handle different situations. At that point I was still living with my family, so I couldn't escape my circumstances. I had a hard time trusting people and was very guarded. I determined to fix the problem myself.

My addiction to self-harm had affected every relationship in my life. When I was with my friends or family and needed to cut, I would get agitated, frustrated, irritated, impatient, and angry, looking for an opportunity to be alone. I distanced myself from everyone and would isolate myself. I wanted to be alone so I could focus on cutting. The process took time, preparation, anticipation, the act of doing so, then cleaning up, bandaging myself, and making sure I had it all together before I was with them again. My life had turned into one big terrible cycle and I wanted out! I reached out to Mercy Ministries for help and God opened the door for me to enter the program.

While I was at Mercy, I realized that it wasn't only about changing my behaviors—I had to get to the root issues that were causing me to want to cut myself. I became very frustrated knowing that I was going to have to face the painful areas of my life that needed healing. I had to deal with the painful emotions, feelings, and memories from my past that I had spent years trying to avoid. Facing my problems head on and not running away or trying to escape took a lot of work. I had to look beyond the self-harm to the things that triggered me to want to do it. I had to make up my mind that no matter how hard it got, and no matter how tough the struggle, I was willing to do whatever it took to be free. To be completely honest, that took a lot of humility because it was something I could not do on my own. I had to reach out to God and the people He had placed around me. I also had to be vulnerable and honest.

Learning how to have a real relationship with Christ has opened my eyes to who I really am. I know that my identity does not come from my past experiences or who others want me to be, but in who God says I am. I found so many scriptures that demonstrate and describe God's love and I received a personal revelation of His love for me. Knowing that I am loved, cherished, wanted, valuable, and fearfully and wonderfully made, is what gives me the courage, boldness, and tenacity to fight.

Speaking God's Word out loud also became an important part of my daily life as the truth began to expose the lies I believed. As I heard myself speak the truth out loud, faith started to rise up within me to believe it. My emotions were based on so many lies that it was important for me to renew my mind to God's Word so that I would actually believe it and my emotions would line up with the truth. I set aside time to do this every day because having structure doesn't leave much room for my feelings. A huge revelation for me was that I am to live my life based on truth, not controlled by my feelings.

Healing and freedom came as I was able to truly experience the love and kindness of God. After I graduated from Mercy, I got connected in a church and God began to use other people to demonstrate His love for me. People would come up to me and tell me that I was important to them and encourage me. I feared that other people would hold my past against me, but instead, I have been surrounded with people who are willing to look beyond my past and see me as a new creation in Christ. God has brought people into my life that I have been able to trust, and through them, I have learned that if people can show me love and forgiveness like this, then how much greater is the love and forgiveness of God towards me?

I know that my strength comes from the Lord. Through the strength that God gives me, I can continue to walk in freedom, forgetting what is behind, because I know God has forgiven me. He is a forgiving God—it was forgiving myself that was hard. Recently, I have had a revelation that God does not hold our past mistakes against us once we have asked for forgiveness, so we shouldn't hold them against ourselves. In order to receive God's forgiveness I had to forgive myself. This is not something I only had to do once. I have learned to live daily with an attitude of forgiveness, being quick to forgive others as well as myself.

I thought since I had graduated from Mercy, that it meant I had my life together. I believed that if I even struggled one bit, I had failed, and the last year of my life was a waste. Obviously, this was a lie. The key is that I have to be real and honest with people, and real with myself. I have to surround myself with Godly people, and I can't let myself become isolated. I have to avoid setting myself up to struggle. I also had to learn that God does not want me to walk in perfectionism. Hebrews 4:16 says, "Let us come boldly to the throne of grace, that we may obtain mercy and find grace to help in time of need." God understands our humanity and does not expect us to be perfect—knowing this brings peace and an ability to enjoy life.

There is hope. Dealing with feelings and emotions may be scary, but they are there for a reason. God gave them to us to signal that something is going on inside of us that needs to be addressed. I have a heart to reach the hurting. I want people to experience the freedom and victory that I have experienced. It begins by opening your heart to God and others who want to love you. I did, and I am forever changed.

I am going back to school to finish my degree in psychology to become a counselor. I want people to experience the true love, joy, and peace that are found in Christ. I want to speak to people and share what God has done in my life. Mark 16:15 says, "He said to them, 'Go into all the world and preach the gospel to every creature.'" That's what I'm going to do!

—**Tiffany**

Personal Study Guide—How does this apply to my life?

1. To Tiffany, emotions were very unpleasant and she was desperate to escape them. What feelings or emotions do you tend to avoid?

____

____

____

____

____

2. What situations or people in your life trigger those emotions?

____

____

____

____

____

3. When you turn to God with your painful emotions, you will receive a permanent solution instead of a temporary fix. What exchange does God promise in Matthew 11:28–30?

____

____

4. Part of understanding how to rest is simply understanding who God really is and allow Him to guide you. Read Psalm 23 and write out what this passage reveals about who God is.

   Psalm 46:10 instructs us to be still so that we can rest knowing that God is in control. Take time to do this as you read Isaiah 42:5.

5. God's love, grace, and forgiveness are always available, but you have to receive these things for them to affect your life. This involves being honest, open, and vulnerable about what is going on in your life and emotions, and willing to make the choice to turn from your desire to be self-destructive. What steps can you take today to be emotionally honest with God and with others so that you can experience true love, forgiveness, and grace?

6. Sometimes the only way out is to go through and deal with the painful emotions that you may have been trying to avoid. It may seem hard in the moment, but it is the only way to get beyond your

past to become the person that God created you to be. What things do you find in James 1:2 that God desires to develop in your life as you press through and fully process difficult situations?

____

____

____

____

____

Talk to God

Use this space to write down any prayers, thoughts, or feelings you may have. This is a place to journal how you really feel.

____

____

____

____

____

____

____

____

____

____

Scripture to Study:

"Our God is a God who saves; from the Sovereign LORD comes escape from death."

—Psalm 68:20

Chapter Eight

VALORIE'S STORY: A VICIOUS CYCLE

My dad was a chaplain in the military, so we moved all the time and I had difficulty adjusting and making new friends. I was also a bit heavyset, which made it easier for other kids to find something to tease me about. They would call me fat, ugly, and say mean things like, "Valorie needs to watch her calories!" Over time, I thought if so many different people said all of the same things, their words *must* be true.

I grew up in church, but most of the time I attended military style chapels. Although I was on a worship team, it was always a performance for me. I wanted my time in the spotlight—my 15 minutes of fame. Even though I loved to sing, what I really wanted was to perform and get attention, and my heart was never in the right place. Since we were always moving from church to church, I never felt like I was growing spiritually because I was not able to really be connected. Slowly I built up resentment toward God for wasting my time and bouncing me from place to place, school to school, and church to church. I doubted His love for me and felt rejected by Him, so I strayed away from my relationship with Him.

I believed that no one truly liked me. In fact, I believed so many lies about myself that I thought I was worthless and my life had no purpose. No guy had ever asked me out or even taken an interest in me, so I believed I was unworthy of love and no man would ever marry me. I felt that nobody wanted me, not even God.

I didn't want to show emotions or allow anyone to see me cry in school. I didn't want anyone to know that they were hurting me, so I would self-harm as my secret way of being able to let out my emotions and keep myself from exploding at school. It was a way for me to cope with how I was feeling so no one could see the emotions that I was dealing with on the inside.

At first my self-injury was minor, but over the years it worsened to the point that I had to be repaired with many stitches. I hurt so much inside and my pain had nowhere to go. I had been hospitalized almost forty times, including a specialized self-injury treatment program. At these places I experienced treatment, but not transformation. My parents felt they had tried everything they knew to help me. After I had exhausted all treatment options, my psychiatrist told my parents to put me in a state hospital indefinitely because I was only going to get worse. For more than nine years, I was trapped in the vicious cycle of self-injury. It consumed my life and my identity.

My doctor wanted to give up on me and I wanted to give up on myself because nobody knew how to help me. I was tired of trying and I didn't know what else to do. I felt so spent, and I just didn't want to do it anymore. When my doctor wanted to put me in the state hospital, I was like ok—whatever, but at the same time, I was afraid because I knew that any hope I might have had of being free would be gone. In a panic, I started searching online for hospitals and treatment centers, looking for help.

I found several residential hospitals, but my insurance wouldn't cover residential treatment. It would only cover inpatient acute care. I was so desperate to escape the reality of everything going on around me that I started abusing my anxiety medication and using marijuana. There was so much chaos going on and no one knew what to do with me.

A couple of days later, someone told me about Mercy Ministries in a chat room and I grabbed the link. I kept reading everything on their web site and it sounded too good to be true! The program was free of charge. I kept reading, trying to find the "catch," but couldn't find one! I felt hesitant, but realized I had no other choice but to apply.

When I walked through the doors at Mercy, I felt a difference right away from any other place I'd ever been. It was such a soft warm environment compared to the stiff environments of hospitals and other programs I'd been to. The furniture was brand new and exquisite, and two beautiful miniature collies were there to greet me. The staff was so caring and loving, and always made themselves available to talk and pray with me.

My stay at Mercy wasn't easy though, as I was dealing with some tough issues and working through the painful memories from my past. In the beginning, I simply wasn't serious about my healing and continued to walk in a victim mentality. I believed I was sick and would always be sick. I thought to myself that this would be just another treatment place, and I would still be sick when I left, so I hid small sharp objects in my belongings "just in case" I needed to cut. My counselor kept telling me, "Valorie, you can't hold on to Plan B. God has a plan for your life and it is way better than anything you could plan for yourself."

When I was faced with the realization that hundreds of other girls were waiting to get into the program, I knew that if I was going to stay at Mercy and fight for my freedom, it was time to get serious. I did not have anything to go back to. I had failed at so many other treatment programs before, and my only remaining option was a state hospital. That was when I realized how far down the "rabbit hole" I'd gone. I decided that I didn't want to be sick anymore. I *really* wanted my healing—for good.

This was a *huge* turning point for me, and I started experiencing *so* much freedom, healing, and deliverance! God showed me His amazing and unchanging love. I am more beautiful than rubies, diamonds, or pearls because I am His precious daughter who He handcrafted uniquely. I learned to love spending time in His Word and in prayer. My identity is not in my past because I am a new creation. My identity is in Christ and who I am in Him!

I began to see that I needed to change my attitude about a lot of things, including how I viewed reading the Bible. I had never really spent much time reading God's Word before. I would read a verse here and there, but I never really studied the scriptures in depth. At Mercy, we had time set aside each morning to read a passage and discuss it with a staff member. During that time, I really developed a passion for the Word of God and a desire to know to understand it.

After I graduated, I started setting aside a special time every day to read a small section of the Bible and then journal about it. I would write about what God was showing me from His Word that day, and how it applied to my life. Now, I actually look forward to this special time with the Lord every day and to what He has to say to me!

More than two years after graduating, I am still walking in freedom. I have had times where I've dealt with depression, but I've gotten through them without self-injury or other destructive behaviors. My relationship with God continues to grow stronger and stronger every day as I spend time with Him and actually apply His Word to my life.

I love encouraging girls who are struggling with the issues that I once had. I understand the things that they are going through and want them to know that they can have hope because there is freedom. As I am faithful to walk according to God's Word, He continues to reveal Himself to me more and more.

I recently had the opportunity to audition for the vocal team at my church. I wanted to make sure that my heart was in the right place and that my motives were truly to worship God and not to gain attention by performing. I began to pray about it, and I felt a peace about my decision to audition. I was soon invited to be part of the worship team! It was a really awesome experience for me.

Through it all, I have learned that self-injury is a choice. God says in Deuteronomy 30:19, "Today I have given you the choice between life and death, between blessings and curses. Now I call on heaven and earth to witness the choice you make. Oh, that you would choose life, so that you and your descendants might live!" It's all about choices. As for me, I choose life!

—Valorie

Personal Study Guide—How does this apply to my life?

1. Growing up, Valorie heard many negative and hurtful comments that she accepted as truth. What are some hurtful things you remember being said to you growing up and how have those things affected what you believe about yourself?

___

___

___

2. Valorie's beliefs about herself led her to pull away from relationships and eventually turn to self-harm as a way to express her emotions. How have your beliefs about yourself shown up in your behaviors?

3. Valorie began to resent God after having to move around so much. When have you felt anger or resentment towards God?

4. It is important to admit how you feel, but it is vital to see the truth despite your emotions. What truth do you find about God in Psalm 103? How would your feelings toward God change if you chose to believe the truth?

5. God tells you in Ephesians 4:26 that it is OK to be angry, but not to sin. What are some healthy ways to express your anger that would not be harmful to you or anyone else?

6. Valorie was challenged by her counselor to ask herself how badly she wanted to be free. Write your answer to that same question: How badly do you want to be free?, along with some practical steps that you are willing to take to choose that freedom.

Talk to God

Use this space to write down any prayers, thoughts, or feelings you may have. This is a place to journal how you really feel.

Scripture to Study:

> "Return to the Lord your God, for He is merciful and compassionate, slow to get angry and filled with unfailing love. He is eager to relent and not punish."
>
> —Joel 2:13 (NLT)

ABOUT MERCY MINISTRIES

Mercy Ministries exists to provide opportunities for young women to experience God's unconditional love, forgiveness, and life-transforming power. We provide residential programs free of charge to young women ages 13–28 who are dealing with life-controlling issues such as eating disorders, self-harm, addictions, sexual abuse, unplanned pregnancy, and depression. Our approach addresses the underlying roots of these issues by addressing the whole person—spiritual, physical, and emotional—and produces more than just changed behavior; the Mercy Ministries program changes hearts and stops destructive cycles.

Founded in 1983 by Nancy Alcorn, Mercy Ministries currently operates in three states and in Australia, Canada, New Zealand, and the UK, with plans for additional U.S. and international locations underway. We are blessed to have connecting relationships with many different Christian congregations, but are not affiliated with any church, organization, or denomination. Residents enter Mercy Ministries on a voluntary basis and stay an average of six months. Our program includes life-skills training and educational opportunities that help ensure the success of our graduates. Our goal is for each young woman to not only complete the program, but also to discover the purpose for her life and bring value to her community as a productive citizen.

For more information, visit our Web site at www.mercyministries.com.

Mercy Ministries of America
www.mercyministries.com

Mercy Ministries Australia
www.mercyministries.com.au

Mercy Ministries Canada
www.mercycanada.com

Mercy Ministries UK
www.mercyministries.co.uk

Mercy Ministries New Zealand
www.mercyministries.org.nz

Mercy Ministries Peru
www.mercyministries.com

ABOUT THE AUTHOR

During and after college, Nancy Alcorn, a native Tennessean, spent eight years working for the state of Tennessee at a correctional facility for juvenile delinquent girls and investigating child abuse cases. Working for the state allowed her to experience firsthand the secular programs, which were not producing permanent results exemplified by changed lives. Nancy saw many of the girls pass the age of eighteen and end up in the women's prison system because they never got the real help they needed. She knew lasting change would never come as the result of any government system.

After working for the state, she was appointed Director of Women for Nashville Teen Challenge, where she worked for two years. Through her experience, she came to realize that only Jesus could bring restoration into the lives of these girls who were deeply hurting and desperately searching for something to fill the void they felt in their hearts. She knew God was revealing a destiny that would result in her stepping out to do something to help young women.

In January 1983, determined to establish a program in which lives would truly be transformed, Nancy moved to Monroe, Louisiana, to start Mercy Ministries of America. God instructed Nancy to do three specific things to ensure His blessings on the ministry: (1) not to take any state or federal funding that might limit the freedom to teach Christian principles, (2) to accept girls free of charge, and (3) to always give at least 10 percent of all Mercy Ministries' donations to other Christian organizations and ministries. As Nancy has continued to be faithful to these three principles, God has been faithful to provide for every need of the ministry just as He promised.

In Monroe, Nancy began with a small facility for troubled girls. After adding on twice to make additional space in the original home, Nancy began to see the need for an additional home to meet the special needs of unwed mothers. For this dream to be realized on a debt-free basis, Nancy knew she would need to raise funds. No doubt, God knew the need and already had a plan in place.

One day, Nancy, exhausted from speaking at an evangelism conference in Las Vegas, boarded a plane for home. The man sitting next to her seemed ready for a chat. When he asked her how much money she had lost gambling, Nancy told him she hadn't gone to Vegas to gamble and shared briefly about Mercy Ministries with him. He seemed interested, so Nancy gave him a brochure as they parted. About four weeks later, this same man called Nancy to ask her for more details about Mercy Ministries and said he felt compelled to help in some way. It was then that Nancy told him about the plans for the unwed mothers' home. He told her he had been adopted when he was five days old. His heart was so touched that he wrote a check to Mercy Ministries for the exact amount needed to help build the second Mercy Ministries house debt-free.

You can read Nancy's entire story in her book Echoes of Mercy.